Bizarre Beast Battles

Kangaroo vs. Gorilla

By Janey Levy

Please visit our website, www.garethstevens.com. For a free color catalog of all our high-quality books, call toll free 1-800-542-2595 or fax 1-877-542-2596.

Library of Congress Cataloging-in-Publication Data

Names: Levy, Janey, author.
Title: Kangaroo vs. gorilla / Janey Levy.
Other titles: Kangaroo versus gorilla
Description: New York : Gareth Stevens Publishing, [2022] | Series: Bizarre beast battles | Includes index. | Audience: Grades 2–3
Identifiers: LCCN 2020032578 (print) | LCCN 2020032579 (ebook) | ISBN 9781538264799 (library binding) | ISBN 9781538264775 (paperback) | ISBN 9781538264782 (set) | ISBN 9781538264805 (ebook)
Subjects: LCSH: Kangaroos—Juvenile literature. | Gorilla—Juvenile literature.
Classification: LCC QL737.M35 L495 2022 (print) | LCC QL737.M35 (ebook) | DDC 599.2/22—dc23
LC record available at https://lccn.loc.gov/2020032578
LC ebook record available at https://lccn.loc.gov/2020032579

First Edition

Published in 2022 by
Gareth Stevens Publishing
29 E. 21st Street
New York, NY 10010

Designer: Katelyn E. Reynolds
Editor: Therese Shea

Photo credits: Cover, p. 1 (kangaroo) CraigRJD/iStock/Getty Images Plus; cover, p. 1 (gorilla) Gail Shumway/Photographer's Choice/Getty Images Plus; cover, pp. 1–24 (background texture) Apostrophe/Shutterstock.com; pp. 4–21 (kangaroo icon) DGBULBUL/iStock/Getty Images Plus; pp. 4–21 (gorilla icon) AlonzoDesign/DigitalVision Vectors/Getty Images; p. 4 MrsWilkins/iStock/Getty Images Plus; p. 5 Arterra/Universal Images Group via Getty Images; p. 6 prospective56/iStock/Getty Images Plus; p. 7 Alan Tunnicliffe Photography/Moment/Getty Images; p. 8 Theo Allofs/Corbis Documentary/Getty Images Plus; p. 9 Colin Langford/500px/Getty Images; p. 10 Tier Und Naturfotografie J und C Sohns/Photographer's Choice/Getty Images Plus; p. 11 tirc83/E+/Getty Images; p. 12 Matt Deakin/Moment/Getty Images; p. 13 Andrew Plumptre/Oxford Scientific/Getty Images Plus; p. 14 Imagevixen/RooM/Getty Images; p. 15 Judy Bellah/Lonely Planet Images/Getty Images Plus; p. 16 miralex/E+/Getty Images; p. 17 Anolis01/iStock/Getty Images Plus; p. 18 Australian Scenics/Photolibrary/Getty Images Plus; p. 19 S1001/Shutterstock.com; p. 21 (kangaroo) Freder/ iStock/Getty Images Plus; p. 21 (gorilla) Cesar March/Moment/Getty Images Plus.

Printed in the United States of America

CPSIA compliance information: Batch #CSGS22: For further information contact Gareth Stevens, New York, New York, at 1-800-542-2595.

CONTENTS

Words in the glossary appear in **bold** type the first time they are used in the text.

KICKING KANGAROOS

Do you love kangaroos? They're some of the most popular animals in the world. These Australian herbivores, or plant eaters, often appear in stories and movies.

Kangaroos have a sweet face, large ears, and a long, **muscular** tail. They have small front legs and large, powerful back legs and feet. A female with a baby, or joey, peeking out of her pouch seems cute. The kangaroo's hopping motion is cute too. But don't be fooled. Kangaroos are deadly kickboxing fighters that have even killed people!

Six species, or kinds, of these **mammals** exist. The largest is the red kangaroo.

GREAT GORILLAS

Gorillas are **primates**, just like people are. They can stand up straight, like people, but they usually walk resting part of their weight on the **knuckles** of their hands. They have black skin and hair. Adult males are called silverbacks because of silver hair on their back.

Movies and TV shows sometimes show gorillas as frightening bad guys, but they're actually mostly peaceful. In fact, they eat mainly plants. Males may fight over **mates** or to keep their family safe, however.

There are two species and four **subspecies** of gorillas.

MEASUREMENT MATTERS

Kangaroos and gorillas are both large mammals. Even though they're mostly herbivores, they can be scary. They live in different parts of the world, so they would never meet each other. However, let's see how these two mammals match up! We'll compare size first. For both animals, males are usually bigger than females.

KANGAROO
HEIGHT: UP TO 6 FEET 7 INCHES (2 M) TALL
WEIGHT: UP TO 200 POUNDS (90 KG)

GORILLA

HEIGHT: UP TO 6 FEET (1.8 M) TALL
WEIGHT: UP TO 485 POUNDS (220 KG)

Kangaroos can be a few inches taller than gorillas. But gorillas weigh more than twice as much as kangaroos! Gorillas win this round.

There's a name for the way kangaroos hop, pushing off the ground with both feet at the same time. It's called saltation (sahl-TAY-shun). One kangaroo was recorded covering over 44 feet (13.5 m) in a single hop!

KANGAROO

TOP SPEED: ABOUT 35 MILES (56 KM) PER HOUR

GORILLA

TOP SPEED: ABOUT 20 MILES (32 KM) PER HOUR

The world's fastest human runs about 27 miles (43 km) per hour. When hopping quickly, kangaroos easily beat this speed! How fast are gorillas?

Gorillas would lose a race with the world's fastest human—and with a kangaroo. Kangaroos hop almost twice as fast as gorillas run. Kangaroos win here!

BATTLE BEHAVIORS

How do they compare in battle? Kangaroos fight to **protect** themselves from predators. Males also fight each other to mate with females.

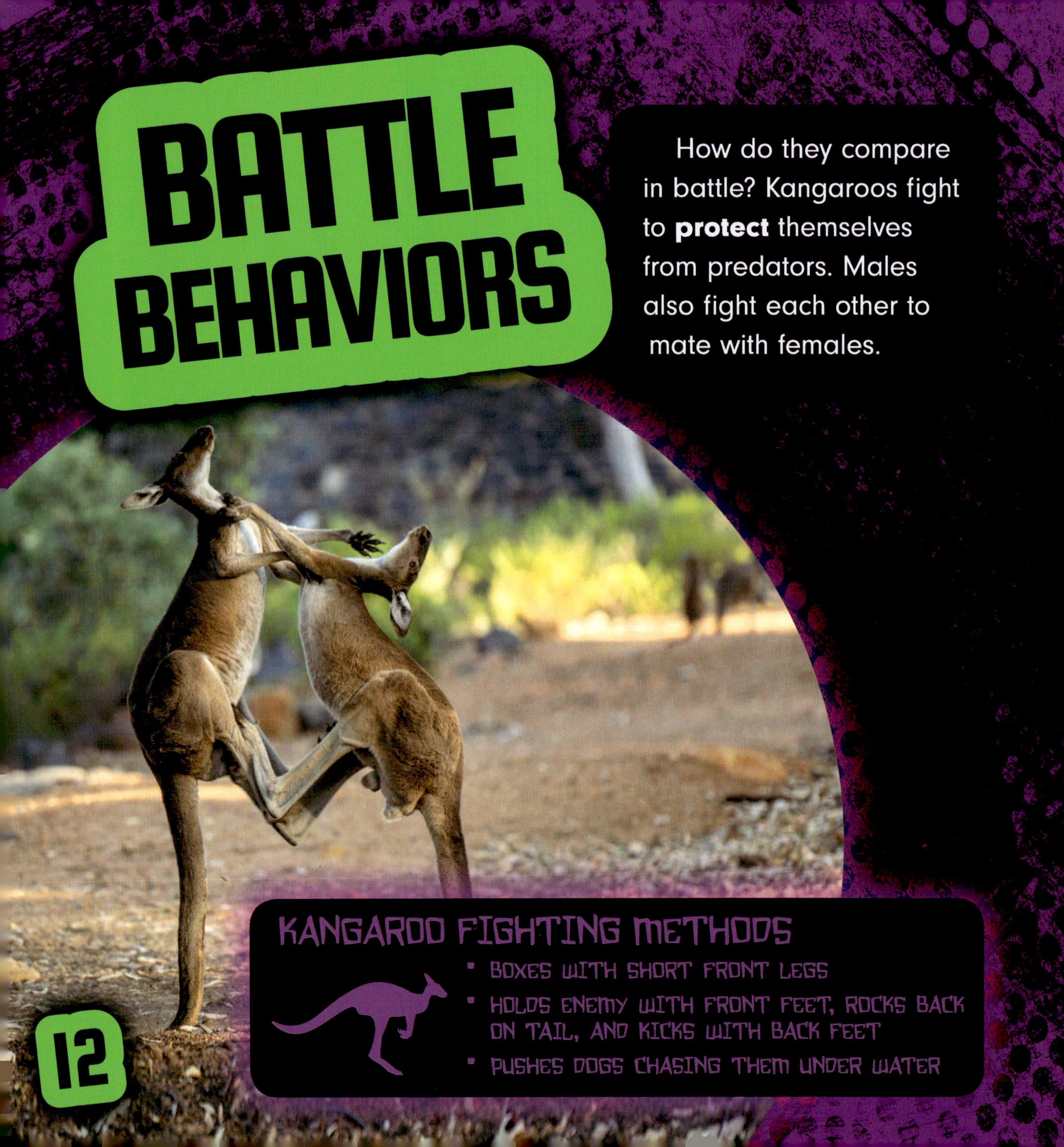

KANGAROO FIGHTING METHODS

- BOXES WITH SHORT FRONT LEGS
- HOLDS ENEMY WITH FRONT FEET, ROCKS BACK ON TAIL, AND KICKS WITH BACK FEET
- PUSHES DOGS CHASING THEM UNDER WATER

GORILLA FIGHTING METHODS

- BEATS CHEST
- ROARS
- CHARGES TOWARD OUTSIDER
- BREAKS OFF BRANCHES AND SHAKES THEM

Gorillas' strength is on display when an outsider shows up. Silverbacks put on an **aggressive** show to scare the outsider and drive them away.

Kangaroos can hurt each other or their enemies. Gorillas usually don't hurt anyone. Kangaroos win this round. They're more likely to be dangerous!

FINDING FOOD

How good is each animal at finding food? Kangaroos eat many kinds of plants. They eat ground plants, such as grass, flowers, and moss. But they also have a way to reach the high tree leaves they like.

KANGAROO FOOD-FINDING METHODS

- USES HANDS TO PULL DOWN TREE BRANCHES IN ORDER TO REACH LEAVES
- TRAVELS LONG WAYS TO FIND FOOD WHEN IT NEEDS TO

GORILLA FOOD-FINDING METHODS

- REMEMBERS WHERE AND WHEN FAVORITE FOOD RIPENS
- USES STICKS TO EAT ANTS TO AVOID GETTING STUNG

Gorillas eat leaves, stems, fruits, seeds, roots, ants, and termites. Adult gorillas can eat up to 66 pounds (30 kg) of food in a single day!

Gorillas use brains *and* tools to help them find food. Does that mean gorillas win this round?

SAY WHAT?

Kangaroos live in groups called mobs. They **communicate** a lot with each other. They make many different sounds for different reasons.

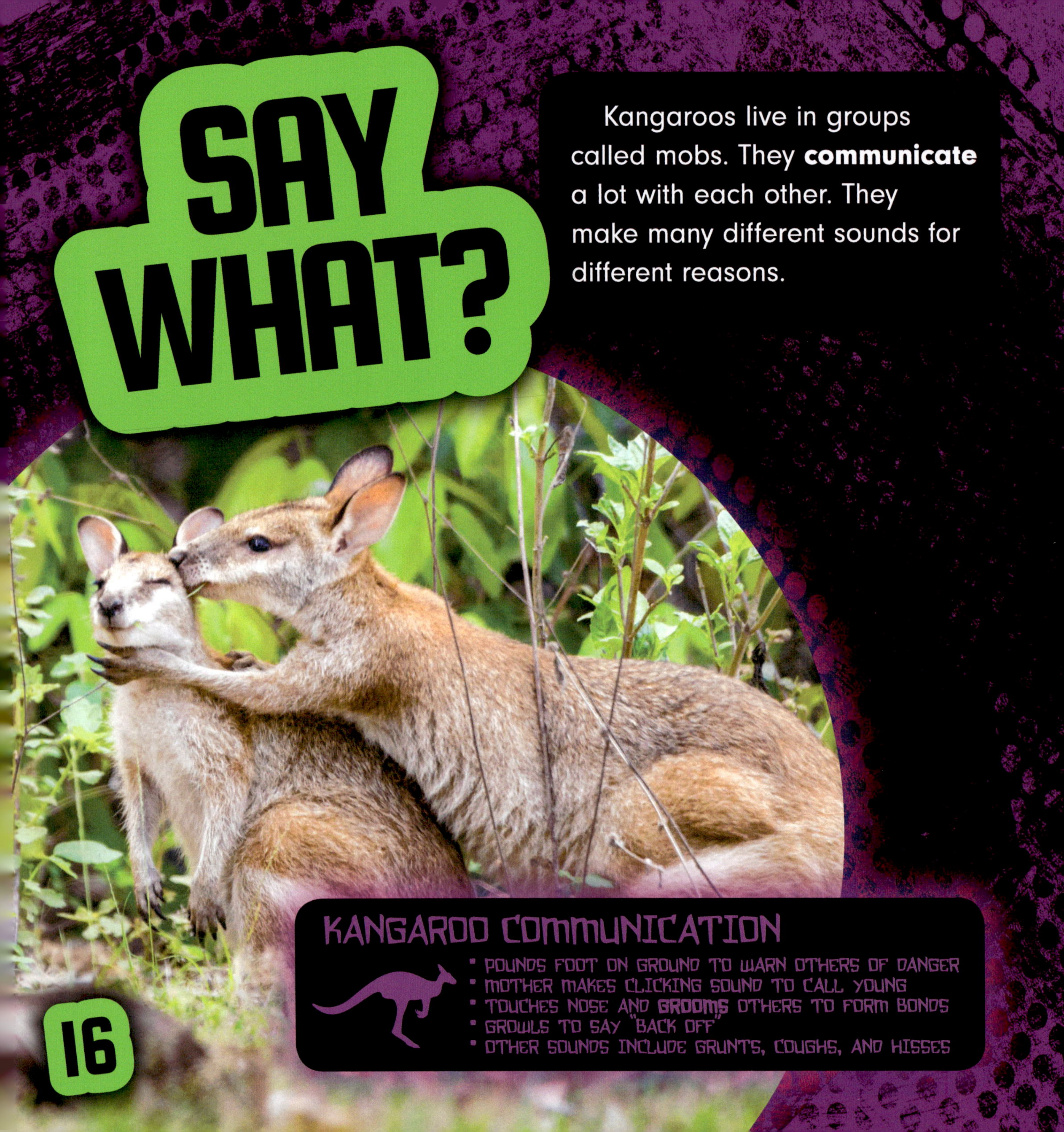

KANGAROO COMMUNICATION

- POUNDS FOOT ON GROUND TO WARN OTHERS OF DANGER
- MOTHER MAKES CLICKING SOUND TO CALL YOUNG
- TOUCHES NOSE AND **GROOMS** OTHERS TO FORM BONDS
- GROWLS TO SAY "BACK OFF"
- OTHER SOUNDS INCLUDE GRUNTS, COUGHS, AND HISSES

GORILLA COMMUNICATION

- SINGS WHEN IT'S HAPPY
- BARKS AT DANGER
- GRUNTS TO TELL OTHERS TO STAY AWAY FROM ITS FOOD
- PARENTS USE TOUCH AND BODY MOVEMENTS TO TEACH YOUNG
- YOUNG LAUGH OR CHUCKLE TO INVITE PLAY
- OTHER SOUNDS INCLUDE SCREAMS, ROARS, GROWLS, HOOTS, AND ALARM CALLS

Gorillas are also social animals and live in family groups called troops. Scientists have found that gorillas make over 20 different sounds. They even laugh!

Both animals have many ways to communicate, with sounds and without. However, gorillas seem to have more than kangaroos. Gorillas win this round!

LIFE SPANS

The six different species of kangaroos have different **life spans**. The red kangaroo has the longest life span. But all kangaroos are in danger because of harm to their **habitat** caused by **climate change**.

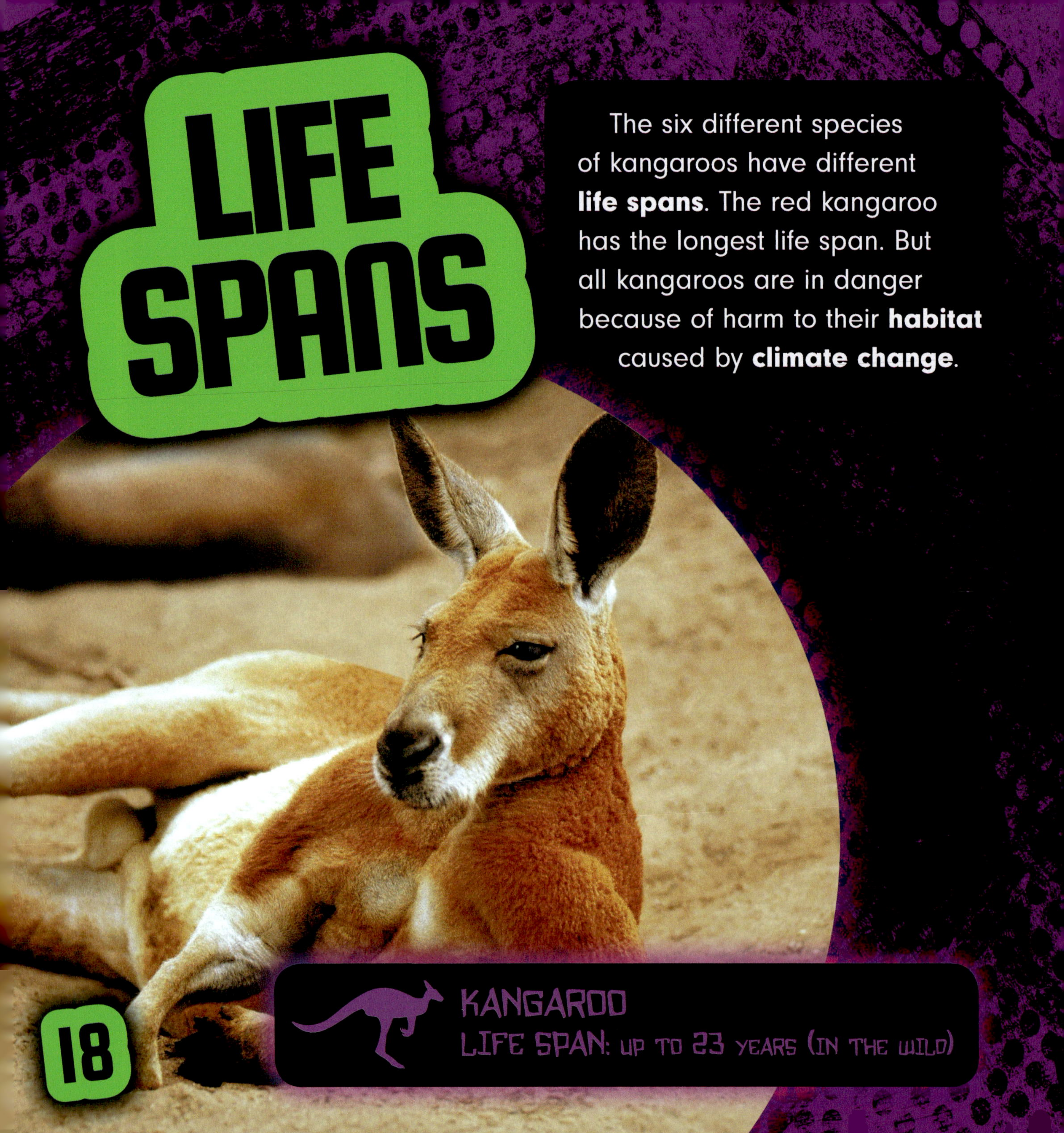

KANGAROO
LIFE SPAN: UP TO 23 YEARS (IN THE WILD)

GORILLA
LIFE SPAN: UP TO 50 YEARS OR MORE (IN THE WILD)

All kinds of gorillas are in danger of disappearing completely in the wild. People are destroying their habitat and hunting them. Yet gorillas can still live long lives.

Gorillas live about twice as long as kangaroos. Clearly, gorillas win this round.

AND THE WINNER IS...

You've learned a lot about these two mammals. Now it's time to decide. Which do you think would win if they battled? Gorillas weigh more than twice as much as the largest kangaroos. And gorillas put on an aggressive display to drive outsiders away. But that display is just a show. Kangaroos cause real harm when they fight.

So it seems as if kangaroos would win. But would a gorilla's aggressive display drive a kangaroo away before a fight took place? You can decide about this beast battle!

KANGAROOS AND GORILLAS LIVE IN DIFFERENT PARTS OF THE WORLD, SO THIS BEAST BATTLE COULD NEVER ACTUALLY HAPPEN. BUT YOU CAN HAVE FUN IMAGINING WHAT MIGHT HAPPEN IF IT DID!

GLOSSARY

aggressive: showing a readiness to attack

climate change: long-term change in Earth's climate, caused partly by human activities such as burning oil and natural gas

communicate: to share ideas and feelings through sounds and motions

groom: to clean

habitat: the natural place where an animal or plant lives

knuckle: a thick, bony body part where the fingers join the hand

life span: the amount of time a person or animal lives

mammal: a warm-blooded animal that has a backbone and hair, breathes air, and feeds milk to its young

mate: one of two animals that come together to produce babies. Also, to come together to produce babies.

muscular: having large muscles, the parts of the body that allow movement

primate: any animal from the group that includes humans, apes, and monkeys

protect: to keep safe

subspecies: a group within a species that is clearly different from others of the species

FOR MORE INFORMATION

BOOKS

Jaycox, Jaclyn. *Gorillas.* North Mankato, MN: Pebble, 2020.

Kenney, Karen Latchana. *Kangaroo Mobs.* Minneapolis, MN: Jump!, 2020.

WEBSITES

Gorilla
www.ducksters.com/animals/gorilla.php
Find out more about gorillas on this site.

Kangaroo
kids.nationalgeographic.com/animals/mammals/kangaroo/
Watch a great video about eastern gray kangaroos.

Red Kangaroo
www.ducksters.com/animals/red_kangaroo.php
Learn more about red kangaroos on this website.

INDEX